T0224071

# Social Infrastructure Maintenance Notebook

Atsushi Yashima · Yu Huang

# Social Infrastructure Maintenance Notebook

 Springer

Atsushi Yashima
Department of Civil Engineering
Gifu University
Gifu, Japan

Yu Huang (ORCID)
Department of Geotechnical Engineering
Tongji University
Shanghai, China

ISBN 978-981-15-8830-3      ISBN 978-981-15-8828-0   (eBook)
https://doi.org/10.1007/978-981-15-8828-0

This Springer imprint is published by the registered company Springer Nature Singapore Pte Ltd.
The registered company address is: 152 Beach Road, #21-01/04 Gateway East, Singapore 189721,
Singapore

# Preface

If we look around us, we can see various lifelines such as roads and railways created with bridges, tunnels, cuts and fills, waterworks and sewage and electricity laid out as a network. These lifelines make up social infrastructures that are essential to our comfortable lifestyles and vibrant economic activities. At the same time, various disaster prevention facilities protecting our lives and properties from earthquakes, cyclones (or typhoons) and rain storms are also extremely important. Do we not take these social infrastructures for granted, thinking "they will be always there, fully functioning"? Have we forgotten that talented engineers are supporting our social infrastructures through their constant efforts?

In Japan where earthquakes, typhoons and rain storms often occur, aging of social infrastructure structures that have been built over the years has become a serious concern. You have probably seen a large-scale repair work being done on an aging structure such as a bridge. Yet because of the constant reduction of public investment and decreasing number of engineers, only strategies our country has come up with are reactive ones, based on ideas such as "we will repair it when it breaks" or "we will repair it after a disaster occurs." As a result, risks of deterioration of national land as well as disaster potential are increasing. In order to build a society that is truly safe and sound, it is "important to make up for engineer shortage and maintain a reasonable balance between new constructions and maintenance/management of existing social infrastructure structures."

In response to such social demands, Gifu University, Japan, established the "Social Infrastructure Maintenance Expert (ME) Training Unit" in FY2008 as part of the "program for creating a hub for developing human resources for regional revitalization" through the Special Coordination Funds for Promoting Science and Technology managed by the Ministry of Education, Culture, Sports, Science and Technology, and we have been conducting activities since then. The unit established the Social Infrastructure Maintenance Expert (ME) Training course, offering a finely tuned curriculum such as lectures including finance, design exercise and field training, so that students can learn maintenance and repair/high functionality technologies for various social infrastructures. The "ME" qualification is granted to engineers who completed this curriculum and passed the accreditation exam.

This book, "Social Infrastructure Maintenance Notebook—Ten Rules for Inspection by Mr. ME," was planned by first graduates who received the ME qualification in FY 2008, with an objective to explain in an easy-to-understand manner the "check" points to keep in mind when inspecting various social infrastructure structures. Instructors who taught the class and led the training worked closely with the MEs and picked ten important points for each type of structures and created a Ten Rules for Inspection. Each rule is accompanied by a brief comment. This book has been put together in a way that not only engineers who are on the front line of maintenance and management but also engineers who are not normally involved in maintenance and management of social infrastructures as well as general public can understand the importance of social infrastructure inspection work.

We would like to express our sincere gratitudes and appreciations to the first MEs who offered their expertise and put together the Ten Rules for Social Infrastructure Inspection, instructors of the Social Infrastructure Maintenance Expert (ME) training course, and everyone who took part in publication of this book.

Finally, during the preparation of English version notebook, special assistance was made by Japan International Cooperation Agency (JICA) and Ms. M. Kumada. Very elaborative draft reading by Prof. K. Sawada, Prof. K. Kobayashi and Prof. S. Murakami is deeply appreciated.

Gifu, Japan                                                                     Atsushi Yashima
Shanghai, China                                                                    Yu Huang

# Contents

# Contributors

| | Items | ME in charge |
|---|---|---|
| 1 | Natural slope | Kouki Nohara |
| 2 | Embankment | Kazuhiro Torimoto |
| 3 | Cut | Koichi Domae |
| 4 | Falling rocks | Toyoshi Kato |
| 5 | Sabo facility | Kentaro Ando |
| 6 | River levee | Tetsuya Ohashi |
| 7 | Retaining wall | Masahiro Suzumura |
| 8 | Snow shed/Rock shed | Takashi Kitade |
| 9 | Tunnel | Yasuaki Okura |
| 10 | Pavement | Yoshinori Katsuyama |
| 11 | Slab | Nobuyuki Soga |
| 12 | Steel bridge | Akihiko Inui |
| 13 | Concrete bridge | Hitoshi Takagi |
| 14 | Box culvert | Eiji Furusawa |
| 15 | Waterworks and sewage | Hajime Koike |

# Chapter 1
# Natural Slope

## 1  Clean Up Slided Soil After Checking the Slope

Slided soil on roads should be removed after confirming the safety of the upper part of the slope, as removal of slided soil on roads could lead to a secondary disaster.

## 2   A Fresh Crack in a Rock Mass is a Warning Sign of a Landslide and/or Failure

If you see a fresh crack in a rock mass, caution should be exercised as it could be a sign of a large-scale landslide and/or failure.

# 3 Carefully Observe Puddles on Roads, Gaps in Pavement, and the Slope

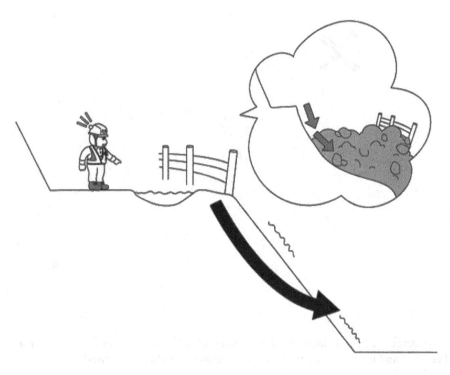

Puddles on roads are sometimes overlooked, however, concavity/convexity and gap of a road surface could be a sign of a landslide and/or failure, so caution should be taken against any changes in the slope.

## 4   A Depression on a Ridge Which Looks Like an Animal Trail is a Warning Sign

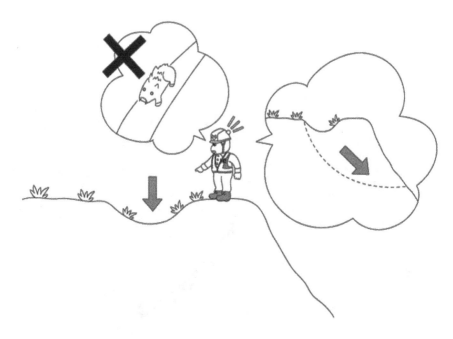

A depression on a ridge could be a trace of a tensile crack at a top part of a landslide and not an animal trail, therefore caution should be exercised.

## 5  Do Not Rest Easy Even If the Rain Stops

Effects of slope variation and water inflow become prominently visible after the rain, therefore a thorough inspection should be conducted.

# 6   Bent Trees are a Warning Sign of a Landslide

Trees with bent bases on a slope could be an evidence that a slide has occurred in the past. Caution should be exercised in future slope inspections.

# 7   If You See Small Fallen Rocks on Roadsides, Pay Attention to the Upper Slope

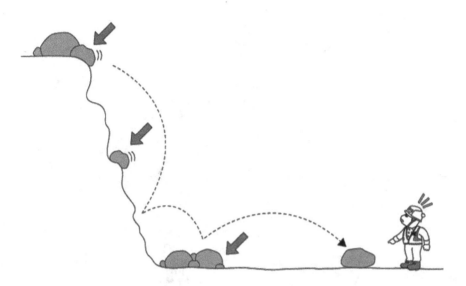

Fallen rocks are often a sign of a failure, therefore the upper slope which is the source of fallen rocks should be inspected.

## 8   Turbid Water at the Foot of a Slope Could be a Warning Sign of a Slope Failure

If turbid water is coming out from the foot of a slope, it could be an indication of a failure at the upper part of the slope, or a sign of a landslide. The slope should be carefully inspected.

# 9   Watch Out for the Valley Side of the Slope as Well

Traces of fallen rocks (such as fallen rocks and damages to paved shoulders) may be seen on a roadside on the opposite side of a slope. Past occurrences of a failure or landslide could be identified by figuring out whether the outcrop on the river bank is a rock mass or earth. Close attention should paid not only on the slope but also roadsides in a broad range.

## 10   Withering of Trees Could be a Warning Sign of a Landslide and/or Failure

If you see withering of trees in a linear form, it could have been caused by a crack on the slope. Close attention should be paid as there is a risk of a landslide and failure.

# Chapter 2
# Embankment

## 1 Watch Out for Cut and Fill Boundary

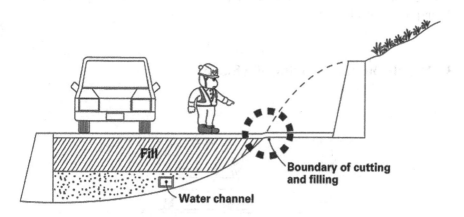

As embankments in a half-bank and half-cut zone are hotbeds of subsidence and slides during earthquakes, extra attention should be paid when inspecting.

## 2   Start with the Top Part of an Embankment Slope

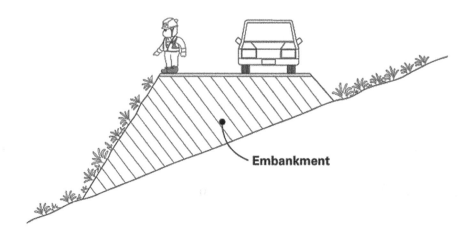

A deformation often appears at the top part of an embankment slope, therefore the inspection should start from there.

## 3   Watch Out for Buckling of a Slope

Buckling of a slope could be an indication of a potentially weakened and deformed embankment due to ingress of water into the embankment.

# 4   A Crack and Deflection of a Road Surface Should be Carefully Observed

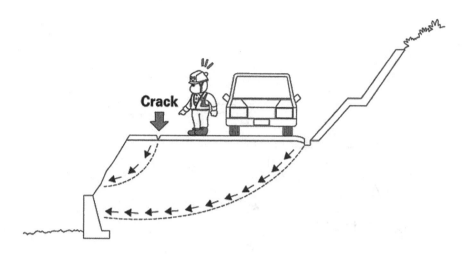

Carefully observe any deformation of a slope and the surrounding conditions which could lead to a failure or cave-in of the embankment. Conduct a simple test such as a two dimensional surface wave exploration as necessary to look for any loosening of the ground.

## 5   Watch Out for Turbid Water as It Could be a Warning Sign of a Slide

If turbid water is observed, it is an indication that a failure is nearing.

# 6   Watch Out for the Level of Water Spring

Water in the embankment should be drained. If the level of water spring indicates the level of groundwater in the embankment, the level could change (increase) due to rainfall and seasonality. Look for any traces of change (traces of water spring and loosened areas of slope) and determine if it effects the stability of the embankment.

# 7  Scour Marks on a Slope Mean Poor Drainage

Scouring of a slope means either the drain is not properly functioning or the drainage facility is insufficient. Saturation from the ground surface could diminish the soundness of the embankment.

# 8 Clogging and Accumulated Water in and/or Around Drain Must be Cleared Immediately

Neglecting clogging and accumulater water blocking a drain is neglecting the management responsibility. Cleanup and other similar measures should be done/taken on the spot. If there is a sediment deposit, a countermeasure should be taken immediately.

# 9   Check the Water Collection into the Drain

Even if the drain is properly functioning, it is meaningless if water does not flow into the drain. It is important to check the drainage when it rains.

# 10   Watch Out for Embankments Facing a River or Seashore

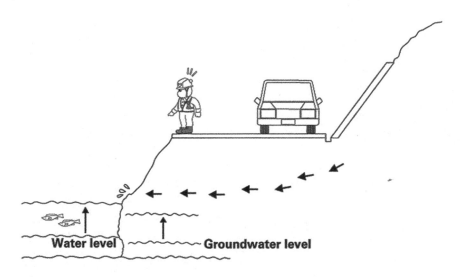

Any place where the variation of groundwater level could effect an embankment, such embankment may be weakened due to a sucking or collapse.

# Chapter 3
# Cut

## 1 Cut Inspection Starts with Your Attire

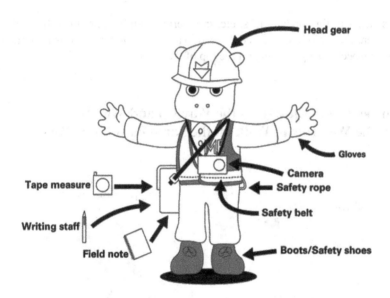

Cut inspection involves risk as footing is often unstable. Make sure to wear safe and comfortable clothing and bring minimum inspection equipment such as a field book, camera and binoculars.

© The Editor(s) (if applicable) and The Author(s), under exclusive license
to Springer Nature Singapore Pte Ltd. 2021
A. Yashima and Y. Huang, *Social Infrastructure Maintenance Notebook*,
https://doi.org/10.1007/978-981-15-8828-0_3

## 2   Walk on the Top of the Slope When Inspecting Cuts

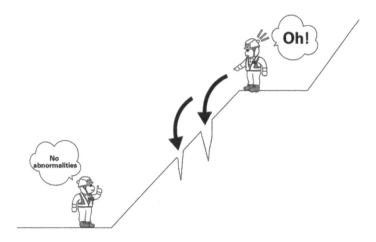

It is easier to find open cracks, etc. in a berm when looking down from above. Cracks near the top of the slope could lead to a large-scale failure. It is important to conduct inspections by walking on the top of the slope.

## 3   Inspect Cuts Closely on the Way Up and Widely on the Way Down. Walk on the Same Path Both Ways

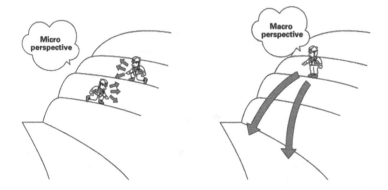

On the way up, get close to small deformations and aging of slope-protection work and inspect with a micro perspective. On the way down, inspect topographic and geological conditions as well as the overall vegetation with a macro perspective. The type and extent of deformation may not be visible if you only conduct an one-way inspection. You could miss a deformation due to factors such as the direction of the sunlight. It is important to walk on the same path both ways.

## 4 Take a Step Further and Inspect the Backland of the Top of the Slope

There may be small swamps (dried swamps) on the backland of the top of the slope. You should also check the scouring and sediment deposit of swamps as swamps could collect a large amount of water and sediment during heavy rains, leading to an unexpected disaster. Also, as the years go by, various conditions may have changed from the original design, and development construction (such as housing development and deforestation) may have been conducted close to the cut surface without your knowledge. Thorough attention should be paid especially on any changes in the catchment area and the drainage capability at the end of the water flow. In addition, the backland of the top part of the slope should be checked as well.

## 5   Watch Out for Cuts Below Fills and Fills Above Cuts

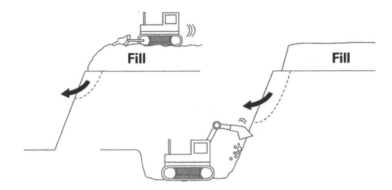

Cuts make slopes unstable because the passive earth pressure decreases. Fills make slopes unstable as they increase the active earth pressure. The combination of these two will double the instability.

## 6   Gully Erosion Leads to a Disaster

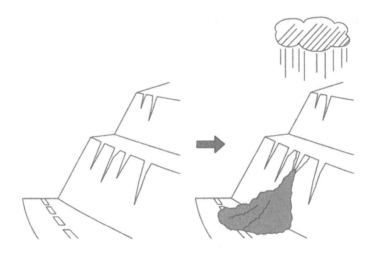

Gully erosion occurs as a result of problematic soil and drainage. If gully erosion on a slope is left untreated, its water channel will expand during a heavy rain and result in a slope failure. It must be fixed quickly.

# 7  If You See Water Spring or a Tensile Crack, Check for Buckling

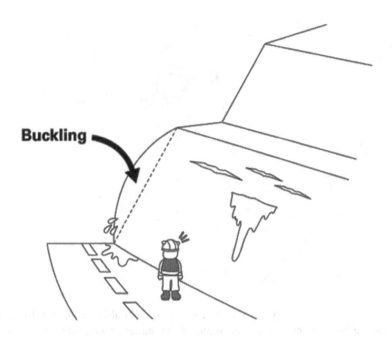

The bottom of a convex on a slope often shows surface water, and the top part of a convex often shows a tensile crack. If you see water spring and a tensile crack, check for buckling.

# 8   Greenery in Winter is a Sign of Danger

Areas with thick greenery have a lot of water. You should be extra careful especially during heavy rains as it can be assumed that either groundwater level is high or groundwater is leaching.

# 9   Watch Out for Deterioration of Slope-Protection Work

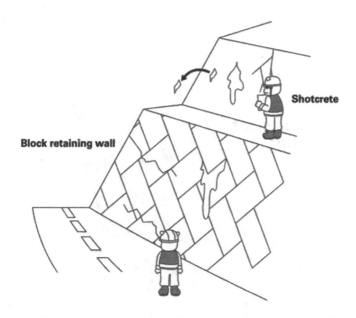

If shotcrete mounted as slope-protection work deteriorates and pieces come off, it is hazardous. You should inspect it by checking the tapping sound with an inspection hammer.

## 10   Be Careful with Cut Slopes Which Failed in the Past

   It is difficult to prevent cut slope failures completely. The cause of the failure needs to be eliminated when restoring the slope. However, in reality, failures often occur at the same place or in adjacent areas. You should remove potential problems with respect to topographic, geological and groundwater conditions for recovery works.

# Chapter 4
# Falling Rocks

## 1 If Small Rocks Start Falling, It is an Indication of a Failure

Before a large-scale rockfall or a slope failure occurs, a small precursory phenomenon such as falling of small rocks may occur. Do not overlook this precursory phenomenon.

A. Yashima and Y. Huang, *Social Infrastructure Maintenance Notebook*, https://doi.org/10.1007/978-981-15-8828-0_4

## 2   Conduct Inspections During Snow Melting Season

Falling rocks in cold regions often occur between 10AM and 12PM during January and April, and they are affected by the rise in temperatures. It is easier to predict risks if inspections are conducted during this season.

# 3  Past Rockfalls are a Sign of Another Rockfall. Falling Rocks Occur Repeatedly

Rockfalls occur during rains, earthquakes and heavy winds, however, it is difficult to predict them because they could also occur after some time. On the other hand, if you understand the characteristics of the soil and geographic composition, you will be able to identify hazardous spots. Knowing that rockfalls occur repeatedly in the same place will lead to disaster prevention.

## 4 Watch Out for Scouring Around Soft Rocks and Boulders

Scouring shows the connection between the flow of water and rockfalls. If the soil on the slope is weak, soil gets eroded and rocks that were buried in the ground start falling.

# 5 Watch Out for Hard Rocks with Open Cracks

Hard rocks/medium-hard rocks with cracks could fall. Observation of cracks is fundamental in figuring out the unsteadiness of loose rocks.

# 6 Distinguish Between an Opposite Slope and a Dip Slope

The direction of rock mass joints and behaviors of failures are related. Knowing that toppling is hard to detect, and that a dip slope could move at anytime while an opposite slope could move suddenly, is helpful in disaster prevention.

## 7   Evidence of Fallen Rocks are Everywhere

It is important to study rocks on roads and traces of fallen rocks. You can estimate the jump distance, falling speed and motion energy.

## 8   Try Looking at the Rock Wall and Slope from a Distance

It is important to observe from a distance. You can identify routes of fallen rocks. There are mass movements which cannot be predicted unless you observe from a distance.

# 9  If You See a Fissure, Conduct a Visual Inspection to Detect Any Deformation

It is extremely important to observe fissures in bonded joints and shotcrete.

## 10   Check the Pockets of Retaining Walls, Fenses and Nets

Check for any changes in soil deposited in pockets behind retaining walls and behind rock nets. It is recommended to remove the soil to the extent possible.

# Chapter 5
# Sabo Facility

## 1 Check the Back Side of Revetments Thoroughly by Walking Along Them

Inspect and see if there are any cave-ins in the ground on the back side of revetments. Sinking of the ground on the back side is caused by soil runoffs, and it poses a risk of collapsing.

A. Yashima and Y. Huang, *Social Infrastructure Maintenance Notebook*, https://doi.org/10.1007/978-981-15-8828-0_5

## 2   Buckling of Revetments is a Warning Sign of a Collapse

Look for any abnormalities in inclination of revetments, such as buckling and falls in the forward direction. Abnormalities in inclination may be an indication of excessive earth pressure and water pressure being placed on the revetment.

## 3  Look at the Riverbed and Make Sure the Foundation is Completely Covered

Check to see if the riverbed is scoured excessively. If the foundation of the revetment is exposed, there is a possibility that the revetment may collapse. If water coming out of the foundation of the dam, there is a risk that the foundation ground may be caving in (hollowing).

# 4  Are Cracks and Water Leaks Signs of Danger?

Check to see if there are any fissures/crack/water leaks in revetments. If there is a water leak, there may be groundwater (a bleeding channel) behind the revetment, adversely impacting the revetment.

# 5 Even If You See No Problems, Tap and Check

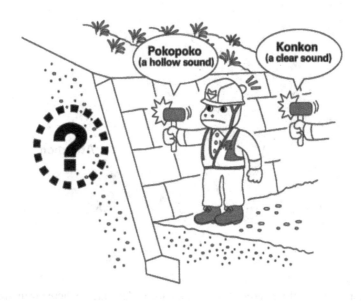

Tap with a wooden hammer, etc. and check for hollowing behind the revetment blocks.

# 6   Drop Structures are Important Checkpoints of Water Flow

Inspect drop structures carefully as they are located in places where the inclination of the riverbed changes, making them prone to deformation. If a drop structure collapses, the riverbed will rapidly deform, significantly impairing the function of channel works. Check for any damages to/peeling of concrete as a large force is being placed on aprons.

# 7 Do Not Overlook Small Deformations. A Large Deformation Starts from a Small Deformation

Look for any fissures/water leaks/missing surface in concrete.

Pay attention even a deformation is small because it could lead to a large deformation. If the dam is made of steel, check for deformation of steel parts (such as dents, bending, ruptures, rust and corrosion).

# 8 Are Joints of the Dam Intact?

Check for misalignment of concrete joints. It is important to determine whether the misalignment occurred during construction or caused by a deformation.

## 9   Is the Dam Reaching Its Limit Due to Excessive Sediment?

Check the sediment behind the dam. Excessive sediment prevents the dam from properly functioning, therefore sediment needs to be removed.

# 10   Surrounding Mountains and Vegetation are Part of Sabo Facilities

Check for natural ground slope failures and landslides, and inspect vegetation around the facility. These symptoms could lead to a large amount of sediment run-off, a change of assumed watershed, or a change of water flow.

# Chapter 6
# River Levee

## 1 When a Flood Occurs, Identify Hazardous Spots

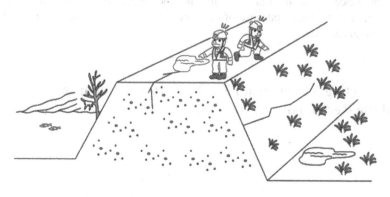

Hazardous spots are easiest to find during and immediately after floods. Finding out which spots are hazardous and in what way plays an important role in effective inspections and flood mitigation activities.

A. Yashima and Y. Huang, *Social Infrastructure Maintenance Notebook*,
https://doi.org/10.1007/978-981-15-8828-0_6

## 2   It is a Good Practice to Conduct Inspections After Grass Cutting

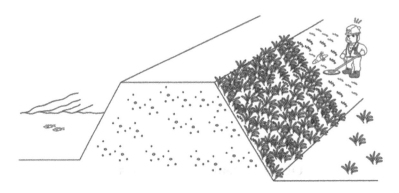

Maintenance of the environment is not the sole purpose of routine grass cutting. Never forget that another purpose of grass cutting is to inspect the levee which is covered in grass, etc.

## 3   Neglecting a Deep Scour Could Lead to a Collapse of the Revetment

If you leave a deep scour untreated, the erosion will progress from the scour, possibly leading to a collapse of the revetment or erosion of the levee.

# 4   Watch Out for Cracks in the Crest

A crack in the crest may be caused by a slide within the levee. Water intrusion into such crack could weaken the levee.

# 5   Cracks and Buckling in a Slope are a Sign of a Slide

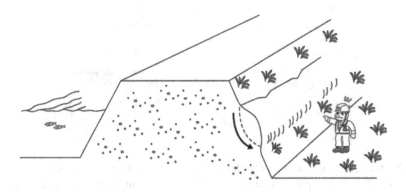

A crack in the slope may be caused by a slide inside the levee. It is important to catch these signs as flooding could further deteriorate a crack or buckling, leading to a levee failure.

## 6   Moisture at the Toe of a Slope May a Signal of Piping

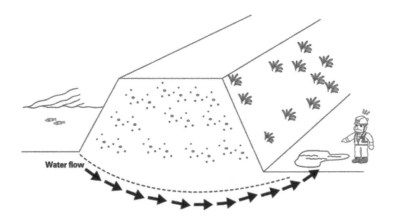

If the toe of a slope is damp, there may be a water channel in the ground. Flooding may induce piping.

## 7   Watch for Boundaries with Structures

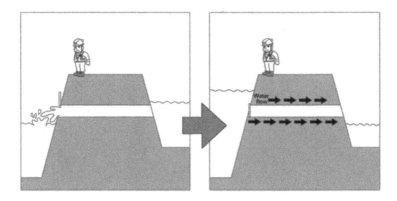

Cavities tend to be formed in the levee in the vicinity of structures such as a sluice gate and sluice pipe, possibly making them a weak point, therefore attention must be paid.

# 8   An Uplift, Cave-In and Sand Boil in a Levee are Signs of a Water Channel

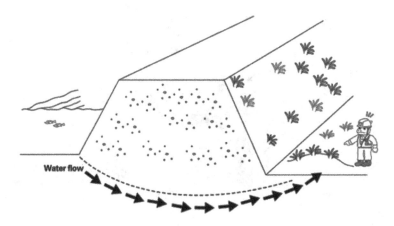

If a water channel is formed in the ground, river water penetrates through the channel during a flood, often causing an uplift, cave-in or sand boil.

# 9   Old Woulds Always Hurt

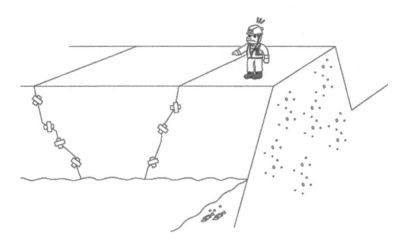

Places that were damaged in the past are subject to a strong external force or have weak ground, and they are likely to cause repeated disasters, therefore you need to keep track of disaster history.

## 10   History Suggests Hazardous Spots

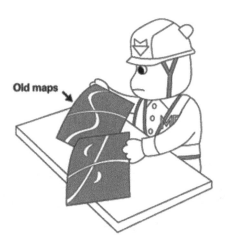

You need to be aware of places that require attention in terms of flood control such as former river channels, slightly elevated area in a former river, existing and former depressions made by overflow, by referring to old topographical maps.

# Chapter 7
# Retaining Wall

## 1 Start the Inspection with the Ground Behind a Retaining Wall

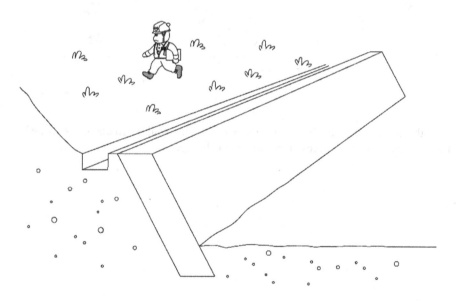

Cracks and sinking/faulting often occur in the ground behind a retaining wall. Inspection must be conducted thoroughly.

## 2   If the Front Side of a Retaining Wall is Excavated, the Wall Will Slide

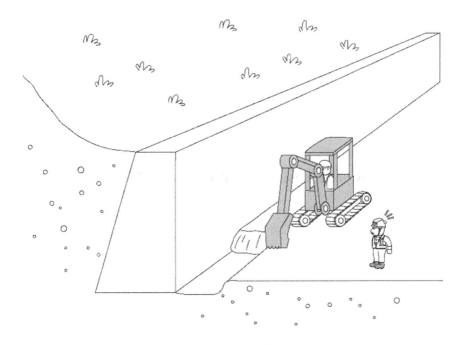

If the ground surface of the front side of a retaining wall is scoured or excavated, the wall may lose passive resistance and start sliding.

# 3  Gaps at Joints Mean a Lack of Bearing Capacity

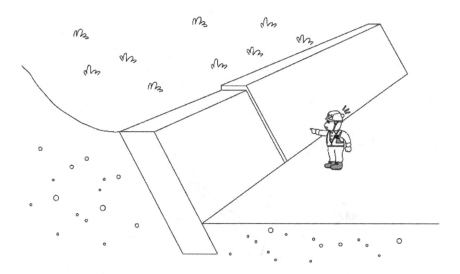

If the bearing capacity is insufficient, misalignments and gaps may occur at joints. You need to examine whether they are progressive, as they may have been produced during construction.

## 4   Buckling of a Retaining Wall Means Excessive Earth Pressure

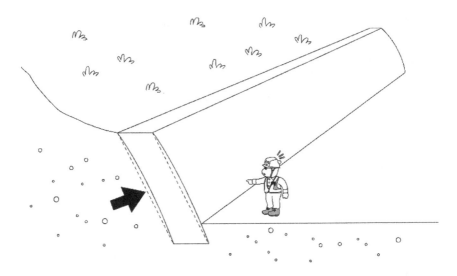

If excessive earth pressure is placed on a concrete block retaining wall, it may result in buckling. If the buckling increases, bonding force of masonry will be lost, possibly resulting in falling and shedding of masonry.

# 5 Insufficient Body-Filling of Masonry Results in Cracks

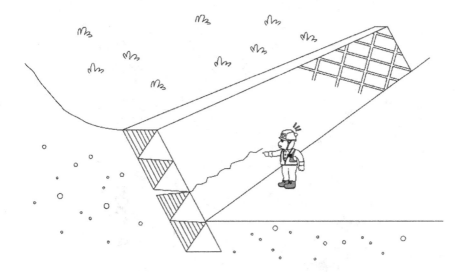

If body-filling of masonry is not sufficient, it may result in cracks.

# 6   Action of Water Pressure is Not Assumed in Retaining Walls

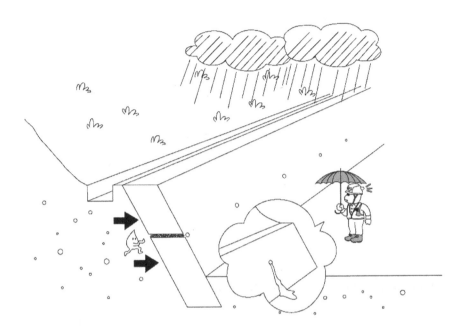

Water pressure is not considered in designing of retaining walls. If water is not properly drained from a drain hole when it rains, excessive water pressure may be placed and the retaining wall may collapse.

# 7   Do Not Let Water in the Backside of the Retaining Wall

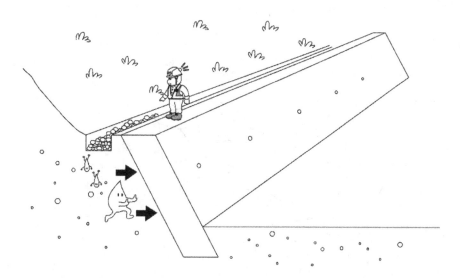

If there is no drain facility in the ground on the backside of the retaining wall, or if there is sediment deposit in a drain, a large volume of water may flow into the backside of the retaining wall, putting excessive water pressure on the wall.

## 8   Caution! Hazards May be Covered with Finishing Mortar

If a retaining wall is repaired with finishing mortar, a new deformation may have occurred on the repaired spot or in the surrounding area.

# 9 If a Sliding Occurs on a Steep Slope, It Will Not Stop

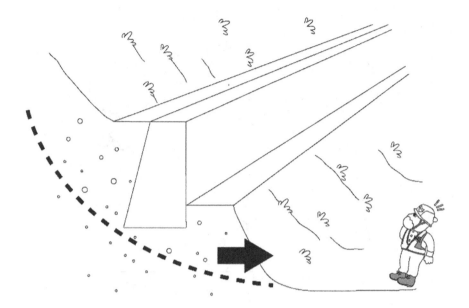

If a retaining wall is built on a steep slope, the foundation ground may cause a circular slip, resulting in a collapse.

## 10   Watch Out for a Fill Behind a Retaining Wall

If there is a fill behind a retaining wall for height raising, caution is needed as unexpected amount of earth pressure may be placed.

# Chapter 8
# Snow Shed/Rock Shed

## 1 Fissures in a Rock Mass or Slope Induce a Failure

Snow sheds/rock sheds are often built on cliffs. If a rock mass or a slope fails, it could lead to a serious accident. It is important to catch signs of rock mass failures.

A. Yashima and Y. Huang, *Social Infrastructure Maintenance Notebook*,
https://doi.org/10.1007/978-981-15-8828-0_8

## 2  Fallen Rocks and Sediment Deposit on a Roof Can Break a Shed

If there are fallen rocks and slided sediment deposit on a roof, structural safety may be insufficient against future occurrences of falling rocks and sediment deposit. Comprehensively determine the situation of the site and remove the deposit as much as possible.

## 3  If There are Any Gaps/Breakage of Roof Materials, Watch Out for a Secondary Damage

If there are any gaps/breakage of roof materials, cushion materials on the roof may fall and cause damages to passing vehicles, etc. In cold regions, water leakage from a gap is hazardous as it may result in falling icicles.

# 4  Watch for Cracks, Delamination and Spalling in Concrete Sheds

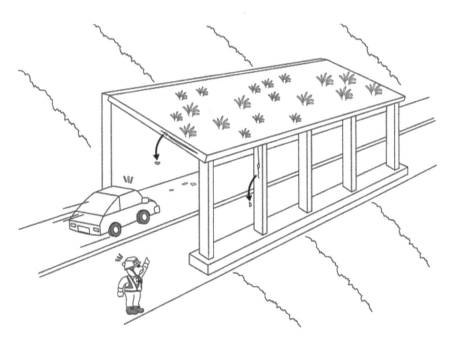

If a shed is made of concrete, it is important to inspect the external appearance and check for cracks, delamination and spalling.

## 5  If a Column is Tilted, Look for Push and Sinking

If a column is tilted, it may have been pushed by increased earth pressure resulting from a landslide or deposition of slided soil, or resulted from sinking, slipping or erosion of the foundation ground on the valley side.

## 6   Check the Base of Columns. Deterioration of Paint on a Steel Shed Causes Corrosion

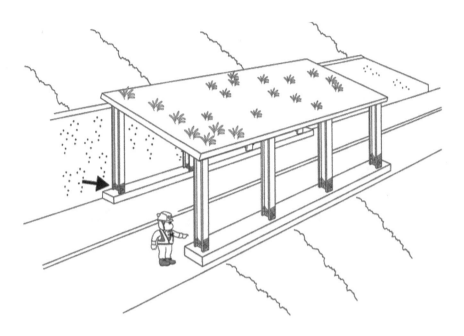

The base of columns in a steel shed are prone to corrode.
This especially applies to cold regions due to spraying of snow-melting agent.

# 7 If a Shed is Made of Steel, Check for Loosening of Turnbuckles

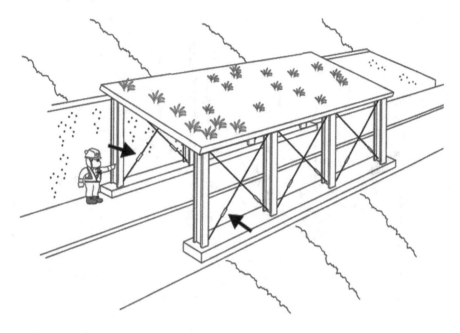

Some steel sheds use turnbuckles as diagonal bracing. Check for any loosening.

# 8   Check the Anchor at the Junction of the Columns and the Foundation

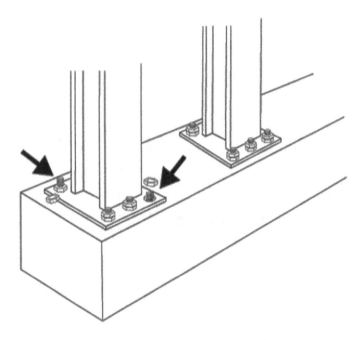

Check to see if there are any deformation, fracture, and rust, etc. on the anchor bolts at the junction of the columns and the foundation.

# 9   Check for Lifting of Foundation and Loosening of Ground by Looking Down

Check and see if the foundation is lifted because of a collapse due to erosion or sinking due to insufficient compaction in the foundation ground on the valley side, etc.

## 10   Jumping Water from Drainage Pipe Should be Prevented Without Stopping the Water

Draining using  drainage pipes is effective in decreasing the water pressure of natural ground.

However, if the water spatters on roads, it sometimes causes cars slip.

# Chapter 9
# Tunnel

## 1 Prepare for Inspection with Existing Documents

In order not miss any defects when inspecting the inside of a dark tunnel within a limited time frame with traffic restrictions in place, it is important to obtain information on the tunnel from existing documents such as design documents and past inspection records. You must thoroughly prepare for the inspection.

## 2   Wear a Mask, Goggles and a Vest During Inspection

A mask, goggles and a vest are the minimum protective equipment to ensure the safety of inspectors. Inspectors must put on a safety harness when conducting any inspection work on a aerial work platform,

# 3 Consider the History of Tunnels When Inspecting Them

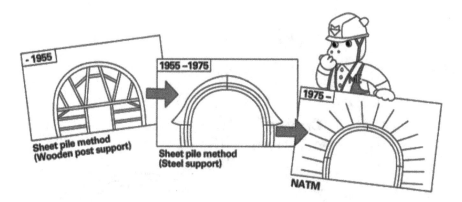

Design and construction methods of tunnels have changed over the years. It is important to find out the characteristics of design and construction methods used when the tunnel was originally built and its repair/reinforcement history, and conduct an inspection with such information in mind and with a focus on places that are prone to defects.

## 4  Identify Cracks with an Understanding of External Force

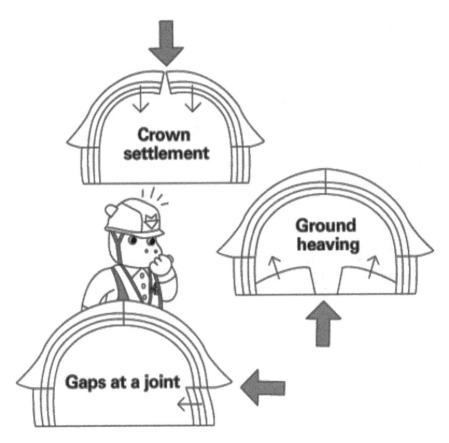

Deformations attributable to external force have different characteristics depending on the cause, such as uneven earth pressure, loosening of ground, insufficient bearing capacity, and water pressure. It is important to understand their characteristics before conducting inspections.

# 5 Start with Susceptible Places When Checking for Spalling

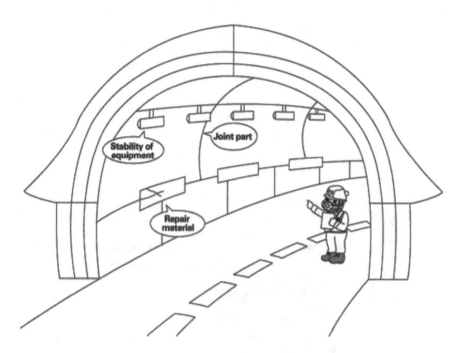

Places susceptible to spalling are joints, repaired areas and where facilities are affixed. It is a good practice to conduct an inspection with a focus on those places.

## 6  Diagnose Delamination/Spalling by the Sound of Tapping Instead of Relying on Your Eyes

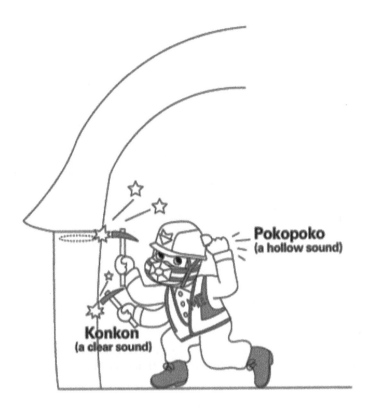

A hammering test is a basic method of identifying delamination/spalling. Do not diagnose a deformation based solely on a visual inspection. It is important to diagnose by comparing the sound of hammer tapping with that of undamaged parts.

# 7    The Amount Water of Leakage Varies Depending on the Timing of Inspection

The amount of water leakage could vary depending on the timing of inspection. Variance of water leakage may be attributable to weather and season in which the inspection is conducted. Even if the amount of water leakage increases, it does not necessary mean that the deformation is progressing. It is important to identify the cause based on the result of routine inspections.

## 8   Look at the Mountain Around the Tunnel Entrance and Check for Slope Deformation and Fallen Rocks

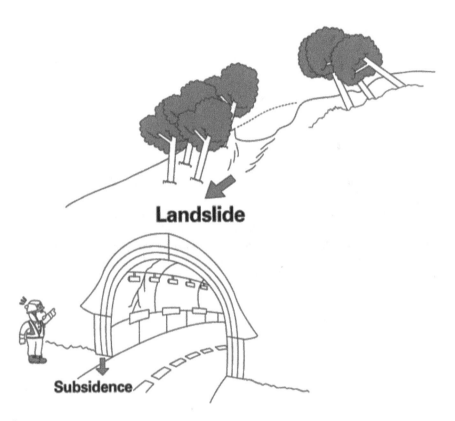

**Landslide**

**Subsidence**

Tunnel entrances are often under shallow overburdens and in unconsolidated ground prone to deformations attributable to rains and earthquakes. If any deformation attributable to external force is found around the lining of a tunnel entrance, it is important to check for any deformation of the surrounding ground, such as a landslide, fallen rocks and sinking of ground.

## 9 Secure Safety with Appropriate Chipping, then Determine If a Temporary Measure is Necessary

If you see delamination/spalling, it is important to secure the safety for the third parties by implementing a temporary measure (knock those parts off by hammering). However, temporary measures that can be implemented during inspections are limited, so the need for setting a wire mesh/net should be determined as necessary.

## 10  Make Sure Debris Do Not Fly into Passersby When Conducting Knocking-Off Work

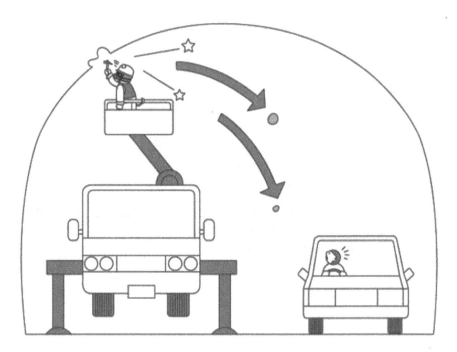

When implementing a temporary measure, it is important to implement safety measures to protect others, such as a measure to prevent debris from flying, deployment of traffic controllers, or implementation of traffic restrictions so others are protected from any accident.

# Chapter 10
# Pavement

## 1 Repair Potholes at Soonest to Ensure Safety and Reassurance

If potholes are left untreated, they could induce an accident which could lead to a defect in the administration of the pavement. They can also damage beneath the pavement surface, therefore they need to be treated at soonest.

A. Yashima and Y. Huang, *Social Infrastructure Maintenance Notebook*, https://doi.org/10.1007/978-981-15-8828-0_10

## 2   Check the Road Surface from the Viewpoint of Motorcycle Driver

You can miss ruts and faults if you are in a car, but you can catch them if you are on a motorcycle.

# 3 Don't Run. Walk Slow and Observe

By walking slow, you may find not only problems in sidewalks but also problems in roads.

## 4   Check the Water Flow and Puddles When It Rains

By conducting an inspection on a rainy day, you can check the water flow and puddles. By doing so, you can identify situations that are obstructing vehicles and pedestrians. It is also useful in checking the effectiveness of drainage pavement.

# 5   If You See a Longitudinal Crack, It May be Caused by Heavy Vehicles or Buried Objects

You must be careful if you see a longitudinal crack. Buried objects are often damaged in such cases.

## 6    If You See a Spider Crack, the Pavement is Probably
at the End of Its Life. It Must be Replaced
from the Bottom

A spider crack is an indication that the pavement is at the end of its life. Replacement construction from the lower part is necessary.

# 7 Check for Gaps When an Unloaded Truck Passes

Small gaps between the pavement and expansion apparatus or a gap in the pavement itself are hard to find. You can check them by listening to the sound of an unloaded truck passing.

## 8   Watch Out for Gaps at Entry Points of Sidewalks

Elderly pedestrians and bicycles could trip even if a gap is small. Entry points and edges of crossing roads tend to have gaps.

# 9   The Crack on Road Surface is a Warning Sign of Deck Slab

If there are cracks on road surface, water may be penetrating into deck slab from the crack. This is undesirable even if the slab is waterproof.

## 10   Do Not Immerse Pavement in Water, Drain Water as Much as Possible

If a pavement is constantly submerged in water, attention should be paid as draining gradient may be unsuitable or a side ditch may be clogged with trash.

# Chapter 11
# Slab

## 1 Defects in a Pavement May be a Sign of Slab Deterioration

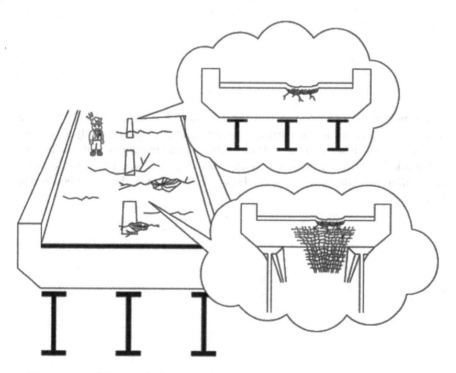

If you see a crack or pothole in a pavement, the beneath slab may be deteriorating, such as a crumbling of the top surface of a slab, cracks in bottom surface of a slab, or slab concrete falling out, etc. If you see any defects in a pavement, inspect the bottom surface of the slab.

A. Yashima and Y. Huang, *Social Infrastructure Maintenance Notebook*,
https://doi.org/10.1007/978-981-15-8828-0_11

## 2  Don't Forget to Check the Ends of Beams for Cracks and Water Leakage

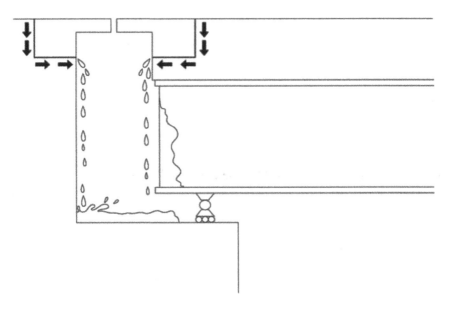

The ends of beams have joints between expansion devices and the slab, therefore susceptible to cracks and water leakage. It is important to inspect not only the bottom surface of the slab but also the ends of the beams (the bottom side of side beams) and check for any deterioration. If water is leaking, it is also important to check the condition of other structural members (such as main girders, side beams and bearing).

# 3    Drainage Failure Accelerates the Deterioration

If a draining facility is clogged, water remains on the road surface, making it easier for water to enter into slab cracks. An effort to remove trash and sediment from the catch basin will help the bridge last longer.

## 4    A White-Colored Crack is a Sign of Deterioration, a Red-Colored Crack is a Symptom of Spalling

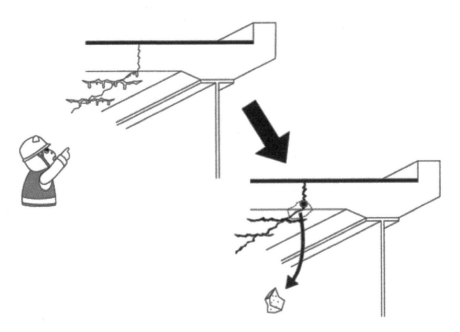

If you see a white-colored crack (i.e. a crack with efflorescence), it means the crack has penetrated through the slab and water is permiating through the crack. It is a warning sign of accelerated deterioration. If you see a red-colored crack (i.e. a crack with rust stain), it indicates that reinforcing steel inside the slab has corroded. As a result, reinforcing steel could swell due to corrosion, causing the concrete cover to spall. Inspections must be conducted carefully especially at places where an accident may harm a third party.

# 5  Pieces of Concrete Could Fall from the Vicinity of V-Cut Drip Channel

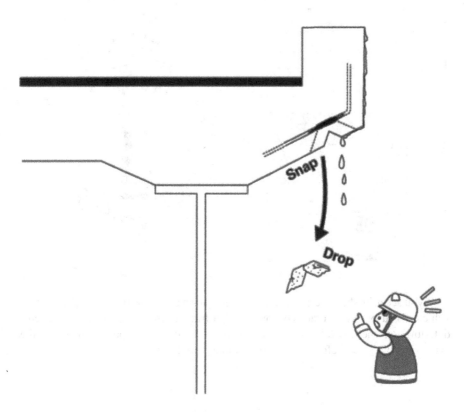

Cover depth of concrete with V-cut drip channel, which prevents water from reaching beams, makes concrete cover depth small, interior reinforcing steel could corrodes due to cracks and carbonation of concrete, making them prone to concrete spalling. Inspections must be conducted carefully especially at places where an accident may harm a third party.

## 6   Grid-Shaped Cracks are Sign of Fatigue

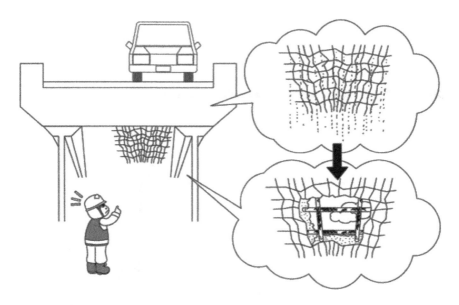

If you see grid-shaped cracks, fatigue degradation due to cyclic loading by passing vehicles is suspected. If cracks become further fragmented, it may result in a severer deterioration such as fallout of a slab concrete, etc. It is important to grasp the symptoms of cracks before the deterioration advances.

# 7 If the Edges of Crack are Missing, It May be a Sign of Accelerated Deterioration

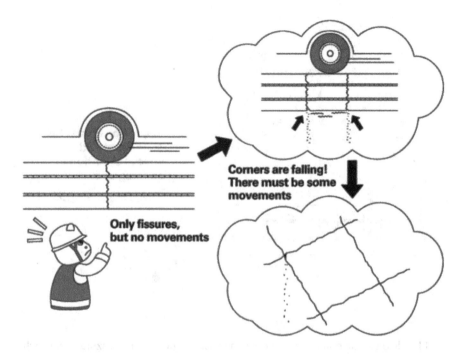

Only fissures, but no movements

Corners are falling! There must be some movements

If grid-shaped cracks are fragmented and their edges are missing, the grid-shaped concrete could rapidly deteriorate and the slab could fall out. Therefore, it is important to conduct up-close inspections in order to find missing edges of cracks. If edges are missing, countermeasures must be taken promptly.

## 8  Corrosion of Reinforcement Steel is Caused by Chloride. Where Does the Chloride Come from?

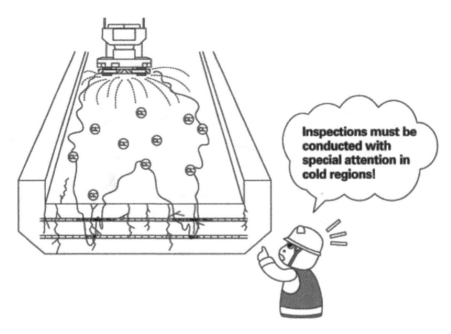

Chloride on a slab comes from the sea in coastal regions, anti-freezing agent such as calcium chloride and sodium chloride used in snowing/freezing weather in mountain regions. Inspections must be conducted with special attention in these regions as chloride ion accelerates the corrosion of reinforcing steel, and countermeasures must be taken promptly to prevent interior reinforcing steel from corroding.

# 9 Trace Back "Water Channels" and Check for Any Damages

Damages to a bridge are often caused by water. It is especially important to find damages to the main structure such as beams by tracing back water leaks and "water channels." At the same time, it is also important to look for "water channels" that are causing the damages to the main structure, by tracing them back from the damages.

## 10   If There are Any Partially-Reinforced Parts, Inspect at the Adjacent Parts Carefully

If a slab is partially reinforced with steel plate bonding, etc., deterioration tends to become visible in adjacent parts that are not reinforced. Therefore it is important to inspect the adjacent parts carefully.

# Chapter 12
# Steel Bridge

## 1 Take a Step Back, Obtain a Whole View

Check the bridges with their whole views, you can find abnormal deflection and deformation of girders. In case of finding any misalignments on the line of wheel guard and railing, the supporting or bearing of superstructure may be damaged.

A. Yashima and Y. Huang, *Social Infrastructure Maintenance Notebook*, https://doi.org/10.1007/978-981-15-8828-0_12

## 2  Feel Bridge Vibration, Find Fatigue Risk

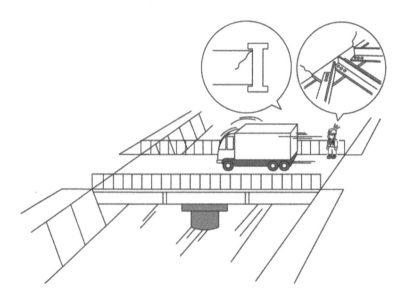

On the "narrow" bridge, which was designed by the economical method, you may feel vibration when heavy vehicles pass. The inspector must pay attention to fatigue cracks of steel girder.

# 3 Behind Abnormal Sounds, There are Something Abnormal in Bridge

Abnormal sounds may be coming from dislocated draining pipes and members in steel girder or damages to expansion apparatus. If you hear any abnormal sound, it is important to inspect the source of sound and identify damaged areas.

## 4   Water Leakage and Deposition of Sediment are a Cause of Corrosion

If water is leaking from a slab or there is a sediment deposit on the bridge seat, it could cause corrosion of steel girder. In bridge maintenance work, it becomes important to eliminate damage factors and avoid formulation of corrosive environment. In the maintenance and management of bridges, it is important to eliminate these causes and avoid creating an environment prone to corrosion.

# 5  Find Layered Rust, Remove Them on the Spot

If layered rust occurs on steel girder, remained water inside the layers accelerate corrosion. In case of finding any layered rust within a reachable range, chip off the layered rust with a testing hammer and apply anticorrosive paint on the spot in order to extend the lifetime of the bridge.

## 6   Do Not Miss the Missing Bolts

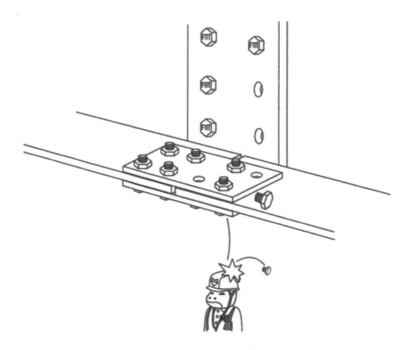

When using high-strength bolts that may hydrogen embrittlement, it is important to check fallen or falling bolts due to delayed fracture. If there is a possibility of third-party damage, a tapping inspection is required.

# 7  If You Find a Gap in Expansion Apparatus at Your Foot, Check the Foot of the Bridge

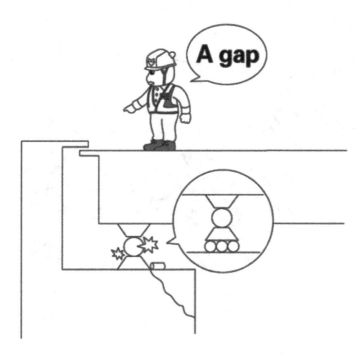

If there is a gap in expansion apparatus, significant failure, such as damage to bearing, buckling of main girder, shear failure of the bridge seat and so on, are concerned.

## 8   The Day After Rainy Day is the Day for Inspection

On a rainy day or the day after rainy day, water leakage from slab and water splashing on steel girder can be observed. It is particularly effective to inspect corroded steel girder on rainy days so that you can identify the water supply.

# 9  Find Water Path, Obtain «Fastest Path»

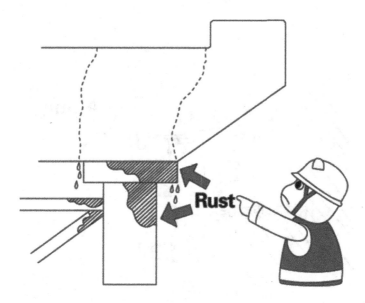

The most important work in bridge maintenance is to block the water. Identifying water path and the treatment against water supply will leads to extend the lifetime of bridge structure.

## 10   Aging Always Start from Painting

The role of painting is to prevent steel girder from corrosion. Steel girder will corrode if the painting deteriorate and does not perform its function sufficiently. In order to extend the lifetime of steel girder, it becomes important not to overlook any sign of deterioration of painting and to take promptly action such as a repainting.

# Chapter 13
# Concrete Bridge

## 1 Cracks in a Pavement will Deteriorates the Slab

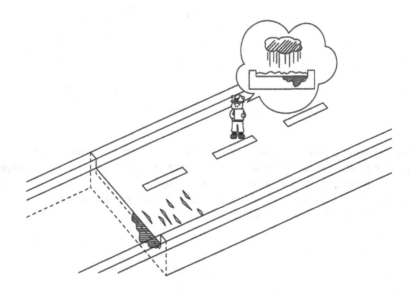

If there are any cracks at places where heavy vehicles pass by (ruts), pay extra attention as rainwater could reach the slab, decreasing the slab's fatigue durability. If there are any cracks in a pavement, record their locations and check if there are any effects of water leakage on the underside of the bridge just beneath the pavement cracks, so that it will help to find any deformation.

## 2   If You See Any Unusual Spacing of Expansion Devices, Pay Attention to the Substructure

Impacts from passing vehicles, expansion and contraction of beams caused by temperature variation, and inclination and shifting of substructure could damage expansion devices or cause abnormalities in spacing between them. If an expansion devices does not work as designed, an unexpected force will act on the slab and/or beams, which could cause a problem.

# 3   Ineffective Draining Will Accelerate the Corrosion of Reinforcing Steel

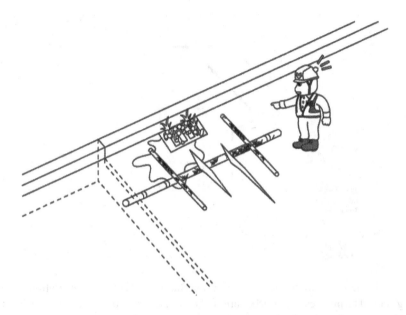

If a drainage system for rainwater, etc. on the bridge surface is not working, reinforcing steel could corrodes at places where water remains or flow down. These should be kept in mind when inspecting the underside of the slab.

## 4   Watch Out for Damages on the Underside of Beams Even If the Bridge Surface Looks Good

Even if there are no deffects on the top surface of a bridge, deterioration of the bridge could be progressing. Carbonation of concrete tends to progress at places with less impact of moisture. Further, initial defects attributable to construction works and corrosion of prestressing tendons attributable to insufficient filling of grout become more dangerous over time. Do not miss any changes in the underside of beams.

# 5   Cracks in Concrete is a Sign of Deformation

Deformation of concrete often starts with cracks. Start an inspection by focusing on cracks. The required detection accuracy should be set according to the purpose of the investigation. At the same time, keep in mind that there are two types of concrete cracks: "non-problematic cracks" and "problematic cracks." It is important to distingish them by guessing the cause.

## 6    Watch Out for  Delamination/Spalling from Cracks

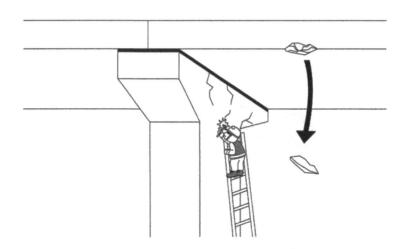

A harmful substance could penetrate into cracks, causing reinforcing steel corro-sion. As a result, the cover concrete could spall off. Delamination tends to occur when there is a sign of water leakage near cracks, or if the cover depth is extremely small, so attention should be paid. Prevention of any damages to a third party is important especially if there are roads crossing below the bridge.

# 7 Check for Corrosion of Reinforcing Steel If You See Cross Sectional Loss of Member

Cross sectional loss often occurs as a result of progression of cracks and spalling, therefore an inspection should also be conducted from a perspective of indentifying the cause that accelerates the corrosion of reinforcing steel.

# 8  Rust Stain from a Filled Section is a Sign of Steel Corrosion

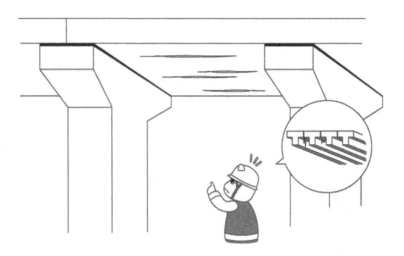

Filled sections of a PC beam are constructed with cast-in-place concrete, and if the concrete is not sufficiently bonded with the joint surface of a PC beam, water drained from the top surface of the bridge may penetrate. In filled sections, transverse prestressing steel are installed perpendicular to the bridge axis in addition to reinforcing steel, therefore attention must be paid as rust stain from filled sections could be caused by corrosion of transverse prestressing steel.

## 9  Water Leakage from the Bridge Surface is Suspected If There is Efflorescence on the Underside of Beams

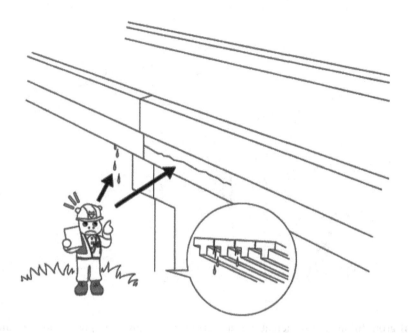

Water leakage from the bridge surface is suspected if cracks on main girders are accompanied by efflorescence. Insufficient grout filling in the main cable duct is also suspected especially if cracks are parallel to the bridge axis and near a steel material. Extra attention should be paid as penetration of rainwater could be related to corrosion of reinforcing steel.

## 10   If There are Any Damages to Anchor Section of Transverse Prestressing Tendon, Check for Steel Corrosion

If grout filling is insufficient in transverse prestressing duct, penetration of rainwater will corrode the tendon, and in an extreme case scenario, PC tendon could fracture or tendon could protrude. Caution is needed especially with a bridge that uses steel rods for transvers prestressing, because in addition to fracture and protrusion, there is a risk that pieces of concrete could scatter and damage a third party. Fractures of transverse prestressing steel rods is difficult to identify visually, but they may be found by finding cracks and spalling of concrete around the anchor section and leaching of rust stain.

# Chapter 14
# Box Culvert

## 1  Pieces of Concrete Always Fall from Above

The highest priority should be placed on prevention of damages to a third party caused by falling pieces of concrete or joint materials, etc. Even if a hammer tapping inspection has been done, new spalling will always occur. Inspections should be conducted assuming that pieces of concrete always fall.

## 2 If You Tap the Surface, You Will Know If There is Delamination

A hammer tapping inspection is effective in finding invisible delamination of concrete. Tap around the area where delamination is suspected, and assess comprehensively by taking into account the result of a visual inspection.

# 3   If Reinforcing Steel is Exposed, then Concrete is Delaminated

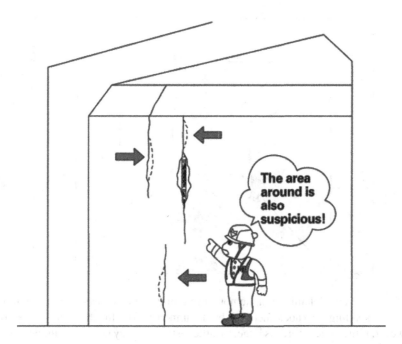

Around the area where reinforcing steel is exposed, corrosion of reinforcing steel inside the concrete considered to be progressing. There is a high possibility that pieces of concrete around the area will spall.

# 4   If You Find Any Delamination, Remove It Before You Leave

If you find any delamination, do not leave it untreated. Always eliminate the risk of pieces spalling by knock them off by a hammer before finishing the inspection. However, if the effected area is large, a traffic restriction may need to be implemented depending on the situation.

# 5 Determine How Old Cracks are by Looking at How Stained They are

Urgency of the need for measures needs to be determined based on when the cracks emerged. If cracks are stained, they are old and are not progressing. If cracks are unstained, they are new and could be progressing.

## 6   Never Fail to Check the Progression of Cracks in Ceiling Board

If cracks on a accelerate is progressing, it is likely caused by a load being applied (a live load, earth pressure). They should be carefully observed as pieces of concrete could fall and damage a third party.

# 7 Is It Sinking? You Will Know If You Look at It from a Distance

If you observe the box as a whole from a distance, sometimes you can find deformations that are hard to find up close, such as openings in joints and cracks in the body that are caused by tilting of the box attributable to differential settlement.

## 8   If There is a Sediment Runoff, Look into Joints to See If There is a Cavity

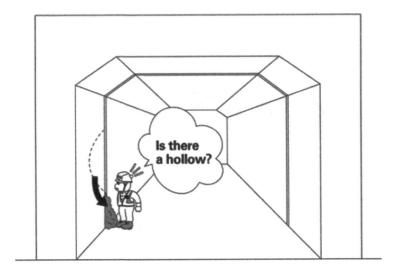

If you see a sediment runoff, it is likely that the backside is hollow, which could cause a deformation of a road surface.

# 9  An Uneven Cover for a Water Channel Creates a Hazard for Pedestrians

Defective cover of a water channel increases the risk of tripping. Although it is not a deformation of a box itself, it should be checked as a focus point to ensure safe use.

## 10   Don't Forget to Check for Deformation of a Road Surface

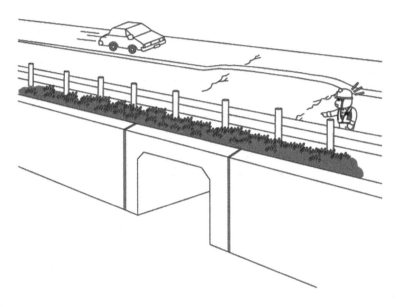

Don't forget to check not only the safety of inner space of the box, but also the safety of the road crossing above the box.

# Chapter 15
# Waterworks and Sewage

## 1 If You Hear a Rattling Sound, Check for Deformation of Manhole Covers

Gata Gata
(a rattling sound)

A rattling sound may be caused by wearing, corrosion and deformation of a manhole cover. It is an indication that the holding frame is separated from the manhole structure and that it is in a hazardous condition. If it is left untreated, a gap will emerge between the holding frame and the surrounding pavement, which may cause skidding, noise and vibration.

## 2   Open the Cover to Check the Flow

Invading tree roots and buildup of mortar and grease block the natural flow, causing bad smell and toxic gas. Dredge and clean as appropriate and ensure the flow capacity of the culvert.

# 3  Check for Bad/Abnormal Smell Coming from the Cover

A toxic gas may be produced inside the manhole.

It is also important to conduct inspections by checking the surrounding environment. For example, check to see if a nearby building pit is beyond its capacity or if city gas is leaking.

# 4   Cracks in Pavement May be Caused by Sinking of the Ground Surface

Check to see if there are any cracks or fissures linearly emerging between manholes. Crack and fissures occur especially in places where a pipe is "slacking."

# 5   Find Damages to the Culvert at an Early Stage by Checking the Inflow of Groundwater

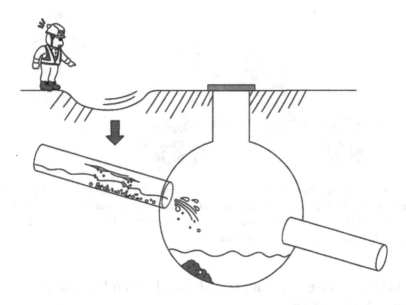

  Culverts may be damaged or cracked due to aging and/or differential settlement as most part of pipeline facilities is buried under roads. Sediment could enter into such cracks and possibly cause a sinking of a road surface.

## 6   Check Facilities Underground During Construction of Objects for Exclusive Use

Various facilities other than sewage (such as waterworks, gas, electricity and telephone facilities) are buried under roads. If another company conducts a construction, be present at the scene of the construction and check the safety of invisible parts to prevent damages from the construction.

## 7   Is Clean Water Coming Out of an Edge of a Pavement or a Side Gutter?

It may be caused by groundwater following a rain, but there is a general possibility that a water pipe may be damaged. Measure the concentration of residual chlorine to determine if the source of the water is waterworks or groundwater.

# 8   Check for Malfunction

If a manhole pump is malfunctioning, sludge will deposit inside the manhole and wastewater will overflow from the opening of a manhole cover.

## 9   Respond Quickly to Complaints from Citizens

A fitting pipe may be blocked or nearby pipeline facilities may be malfunctioning. A prompt inspection and treatment are required.

# 10   Is the Amount of Wastewater Unusually Large Compared to a Day of Good Weather?

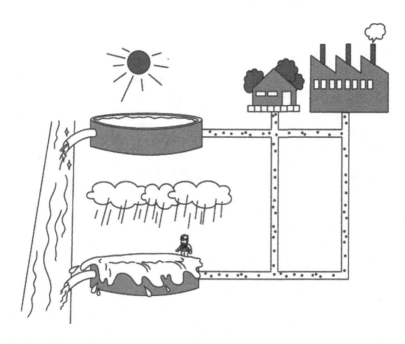

If rainwater is flowing into drainage facilities, the amount of treated water increases. Check to see if households and/or industrial plants are draining rainwater into drainage facilities.

Printed in the United States
by Baker & Taylor Publisher Services